ALEX ALBON

Racing Against the Odds – The Journey of a Formula 1 Star

SCOTT RODMAN

Copyright @ 2024 By Scott Rodman

All rights reserved. No part of this book may be reproduced, distributed, or transmitted in any form or by any means, including photocopying, recording, or other electronic or mechanical methods, without the prior written permission of the publisher, except in the case of brief quotations embodied in critical reviews and specific other noncommercial uses permitted by copyright law.

Contents

INTRODUCTION

CHAPTER 1: ROOTS IN RACING

   Family Influences: The Albon Legacy

   Growing Up Between Cultures: Thailand and the UK

CHAPTER 2: KARTING DREAMS

   Discovering the Thrill of Speed

   Competing at a Young Age

CHAPTER 3: CLIMBING THE JUNIOR RANKS

   Struggles and Breakthroughs in Formula Renault

   Navigating the Competitive World of Formula 3

CHAPTER 4: FORMULA 2 SUCCESS

   Podium Finishes and Career-Defining Wins

   Earning Respect in the Motorsport World

CHAPTER 5: THE CALL TO FORMULA 1

   From Formula 2 to Toro Rosso: The Big Break

   Adapting to the Pace of Formula 1

CHAPTER 6: RED BULL RACING – THE BIG STEP

   Mid-Season Promotion: Joining Red Bull Racing

The Pressure of Competing at the Top Level
CHAPTER 7: UPS AND DOWNS AT RED BULL
Memorable Races and Near Misses
Facing Criticism and Staying Resilient
CHAPTER 8: STEPPING BACK – A YEAR ON THE SIDELINES
Dealing with Disappointment: Losing the Red Bull Seat
Maintaining Focus and Preparing for a Return
CHAPTER 9: THE COMEBACK WITH WILLIAMS
A New Beginning: Signing with Williams F1
Proving Himself Again on the Grid
CHAPTER 10: BEYOND RACING – THE MAN BEHIND THE WHEEL
Exploring Alex Albon's Personal Interests and Hobbies
Honoring His Thai Heritage
CHAPTER 11: LESSONS IN RESILIENCE
Staying Positive in the Face of Adversity
The Importance of Patience and Hard Work

CHAPTER 12: LOOKING FORWARD – THE FUTURE OF ALEX ALBON
   Hopes and Goals for His Formula 1 Career
   The Legacy He's Building in Motorsport
CONCLUSION

INTRODUCTION

Alex Albon's journey into Formula One exemplifies tenacity, determination, and the power of believing in a dream. Albon was born in London on March 23, 1996, and grew up in a multicultural environment that combined his Thai and British ancestry. His Thai mother, Kankamol, and his British father, Nigel Albon, a former racing driver, introduced Alex to motorsport at a young age. Despite having a racing father, young Alex's road to Formula One was not inevitable.

Albon had a lifelong passion for speed. His initial journey into motorsport was through karting, a popular starting point for future racing stars. However, karting was more than a weekend pleasure; it was a serious, competitive, and expensive endeavor. Unlike some of his friends, who affluent families and sponsors supported, Albon's family struggled to meet the sport's enormous financial demands. Racing, with its high costs for equipment, travel, and team support, quickly became a financial strain. Several times, it felt like the voyage might end before it even began.

Despite these obstacles, Albon persevered. He possessed a natural skill behind the wheel, but more than talent was needed in a sport where financial resources frequently decided success. Each race was a challenge on and off the track—finding sponsorships, juggling family finances, and coping with the expectation to succeed. Many young drivers would have given up on their dreams if faced with such insurmountable challenges, but Alex used the adversity to fuel his ambition to achieve.

As he progressed through the karting ranks, Alex encountered increasingly tricky competition. Despite the odds, he began to build a reputation for himself in the racing world, thanks to his innate talent and unwavering dedication. However, not only the physical demands of racing put him to the test but also the emotional toll of knowing that every race could be his last if funds ran out. It was a continuous reminder that his place in motorsport was not guaranteed; he had to earn it daily. Faced with such high stakes from an early age, Albon developed into a driver who thrived under pressure, a trait that would serve him well.

A Dream Worth Fighting For

Albon's journey to popularity was driven by speed and his unwavering belief that his dream of racing in Formula One was achievable, no matter his challenges. By age eight, when many children were still discovering their hobbies, Alex had already engaged in karting. From the start, it was evident that he was a formidable force on the track. As he started winning karting championships, his family's sacrifices and unwavering quest for excellence began to bear fruit. Each triumph in the junior levels brought us closer to our ultimate goal: Formula One.

But as Alex advanced, the stakes rose. The transition from karting to single-seater racing in Formula Renault was a watershed moment, not just in terms of competition but also in terms of financial pressure. In a sport where affluent supporters frequently had the luxury of focussing entirely on performance, Alex had to balance the high stress of racing with the worry of finding enough funding to keep his career going. The line between success and failure was razor-thin, and Alex crossed it daily.

Despite the ongoing uncertainty, Albon's ability shone through. His accomplishments in Formula Renault and, later, Formula 3 drew the attention of those in higher levels of motorsport. Even as his reputation expanded, his financial problems persisted. Alex remained undeterred even when it appeared that his career would stagnate owing to a lack of money. His willpower was tested repeatedly, yet he refused to relinquish his dream.

As Albon advanced through the ranks, his tale became one of endurance. His competitors may have had more accessible pathways, with top-tier equipment and rich sponsorship, but Alex knew he was racing for more than simply victories. He was racing to prove that enthusiasm and hard work can propel you to success no matter where you come from or how the odds are stacked against you.

By the time Alex reached Formula 2, his reputation as a driver who thrived under duress had solidified. He wasn't simply fighting for podiums; he was fighting for the opportunity to continue competing in the sport he loved. Formula One was finally within grasp but remained elusive as financial and competitive demands increased. For many drivers, achieving Formula 2 is a career high point, but for Alex, it was only the next step in a journey that had always been about something more significant. Formula One was more than just a dream for Alex Albon; it was his destiny, which he was willing to battle for at any cost.

And he fought. Every setback, every race that didn't go his way, only added gasoline to his fire. Alex Albon's trip was long, with ups and downs, but he remained steadfast. The odds were stacked against him, but Alex had no doubts. This was a dream worth battling for.

CHAPTER 1: ROOTS IN RACING

Family Influences: The Albon Legacy

In addition to being driven by his love of racing, Alex Albon's journey into motorsport was significantly impacted by his family's history. Alex was accustomed to the sound of engines and the competitive spirit of racing as he was raised in a home where motorsports were already a part of life. In addition to being a dad, his father, Nigel Albon, played a significant role in his early career as a mentor who was aware of the demands of racing.

Nigel Albon has a background in motorsports as well. In the late 1990s, he participated in the British Touring Car Championship (BTCC) and drove touring cars. Even though Nigel never made it to the pinnacles of Formula 1, his involvement in professional racing influenced

Alex's perception of the sport from an early age. Nigel understood more about racing than only the technical aspects of it; he also understood the highs and lows, successes and failures, and sacrifices necessary for success. Alex found great value in this first-hand experience when he started karting and, subsequently, single-seater racing.

When Alex was younger, he would frequently go to the racecourse with his father and take in the sights and sounds that would eventually become a part of his life. In the Albon home, racing was more than simply a pastime; it was a family obsession, and Nigel was crucial in fostering Alex's enthusiasm. But even with his father's history, it was a challenging path. The family faced financial constraints and the high expenses of elite racing, which made each race a significant financial commitment.

However, Alex's destiny was not just formed by Nigel; his mother, Kankamol, was also a significant factor in his development. Kankamol offered the family a unique cultural viewpoint because he is a native of Thailand. In the predominantly Western world of motorsport, Alex's Thai origin became integral to who he was and helped to differentiate him from many of his peers. His mother instilled virtues of perseverance, humility, and hard work—qualities that would serve him well in the often harsh world of Formula 1. His father taught him the technical and competitive aspects of racing.

Despite their disparate cultural upbringings, Nigel and Kankamol shared standard support for Alex's aim. His career was built on his family's persistent belief in his ability. Their impact extended beyond providing emotional support; both parents made sacrifices to help finance Alex's early racing endeavors. The Albons had a strong desire to see Alex succeed, whether through race travel or figuring out how to meet the high financial needs of the sport.

Growing Up Between Cultures: Thailand and the UK

Growing up between two cultures—his mother's Thai origins and his father's British roots—profoundly impacted Alex Albon's childhood. From an early age, Alex's bicultural upbringing broadened his outlook, impacting his life and identity as an athlete. Alex stands out from other racing drivers in motorsports because of his history of navigating two distinct cultures.

Alex, born in London in 1996, was raised surrounded by British culture. His early upbringing was influenced by the fast-paced, competitive lifestyle of the UK, especially in the motorsport community. Thanks to his British father, Nigel Albon, 's strong ties to the industry, Alex was raised with a front-row seat in the racing world. His early years included attending races, speaking English at home, and adjusting to British education.

However, his Thai ancestry had a significant influence in addition to his British upbringing. Kankamol, Alex's mother, saw that he maintained a close relationship with Thailand. Thai culture, which strongly focuses on tradition, family, and respect, permeated every aspect of his life. Alex was introduced to the values, language, and practices of his mother's native Thailand through the family's frequent trips there. He traveled to Thailand to spend time with his extended family, where he was exposed to a different way of life from the fast-paced, individualistic society of the UK.

For Alex, growing up in between these two cultures was a challenge as well as a gift. Being able to represent both of his heritages provided him with a distinct feeling of identity and pride on the one hand. His ability to recognize the variety in the world contributed to his adaptability and receptivity. His capacity for environment navigation would eventually come in handy in the international Formula 1 racing scene, where teams and drivers represent diverse backgrounds.

Living in two different cultures, though, was challenging. At times, juggling his Thai roots and British identity made him feel as though he was torn between two different universes. Because of his Thai ancestry, he was frequently viewed as unique in the UK, but his British upbringing made him strange in Thailand. Due to his inability to completely fit into either culture, Alex had to find his place in both, which led to a complex feeling of belonging.

This cultural diversity affected his perception of the motorsport fraternity early in his racing career. He raced in championships held in the United Kingdom and had a British passport, yet he was also fiercely Thai. Alex's Thai heritage gained prominence as his career developed, especially after he became the first Formula 1 driver to represent Thailand since Prince Bira in the middle of the 20th century. Due to his international success, Thailand's motorsport community saw increased visibility, which sparked interest in the sport in a nation where Formula 1 had not historically garnered much attention.

An essential aspect of Alex's biography is how his bicultural upbringing shaped his identity on and off the track. He carried the principles of family, humility, and respect from his Thai heritage while learning to cherish the discipline and drive instilled in him by his British upbringing. It's a balance that has shaped him into a capable driver who recognizes the value of standing up for something greater than himself. Alex Albon represents how juggling two cultures can be a source of strength, resiliency, and pride as he competes on the global stage.

CHAPTER 2: KARTING DREAMS

Discovering the Thrill of Speed

Alex Albon's deep love for speed and competitiveness led him to embark on an early career in motorsports. He first experienced the exhilaration of racing as a young child, accompanying his father to various motorsport events. He was amazed and excited to see the fast-paced action on the racecourse, and this feeling quickly turned into a strong desire to get involved himself.

He was first introduced to karting when he was eight, which marked a sea change in his racing career. Alex's parents gave him a kart, and he immediately fell in love with the sport. He had never felt anything like the sensation of the wind whipping past him as he raced around the track. Driving gave him an addictive rush of excitement, turning his casual curiosity into a passionate quest. He accepted karting as a necessary component of his life, not just a pastime.

Alex showed an inherent talent for racing when he started to compete in local karting events. He rose through the ranks fast, winning races and becoming known for his abilities. Alex was not scared by the intense competition; on the contrary, he thrived on it. He enjoyed the challenge of testing the limits of his kart and himself, constantly looking for the next thrilling experience. His passion for speed was stoked by every race, which also increased his desire to become a professional motorsports driver.

The karting circuits taught Alex about a lively group of young racers and their goals and objectives. Thanks to their friendship, he grew as a person and competitor. Even in the face of the intense rivalry that is a part of racing, he discovered the value of sportsmanship and cooperation. Mentors and other competitors provided him with support and direction, which helped him improve his track performance. His relationships in those early years set the groundwork for a network of support that would come in handy as his career progressed.

By the time Alex was fifteen, scouts and teams from outside the area had noticed his skill. His remarkable accomplishments in karting championships secured him a position in the fiercely competitive single-seater racing arena. Going from go-karting to cars was a big step, but the speed rush never disappeared. Alex swiftly adjusted, demonstrating his capacity to navigate the intricacies of cornering, braking, and acceleration in a race situation.

He made his single-seater racing debut in the Formula Renault championship, where he went on to do very well. Taking on more experienced drivers just fuelled his ambition to become a Formula 1 driver, the highest level of motorsport. Alex's love of speed increased with every new task, strengthening his dedication to the activity. His motivation shifted to the rush of competing at high speeds, the exhilaration of being on the track, and the sensation of the car's power.

Competing at a Young Age

Ever when he took hold of a go-kart's steering wheel, Alex Albon showed a natural aptitude for racing. He entered the world of competition at age eight, demonstrating a solid grasp of strategy, accuracy, and the art of it all in addition to skill. His early karting track experiences provided a solid basis for his career as a professional driver.

As he rose through the karting levels, Alex encountered many of the usual difficulties confronted by young sportsmen. Many ambitious racers shared the goal of becoming the best driver in the intensely competitive karting environment. Alex loved the rush of competitiveness, even in the face of pressure. He soon gained recognition for his aggressive yet sensible driving style, which frequently helped him stand out from the competition. His ability to remain composed under duress and make snap judgments came in handy when he dealt with difficult racial issues.

Alex started racing in the highly competitive Super 1 National Karting Championship at ten, one of the most challenging events for young drivers in the UK. Alex displayed remarkable consistency, frequently placing on the podium while competing against some of the top young athletes in the nation. In addition to giving him more self-assurance, this achievement attracted the interest of racing organizations searching for the next big thing.

He made considerable advancement in 2010 when he joined Red Bull Racing, a reputable organization committed to developing fresh talent. This collaboration was a turning point in his career. Alex gained improved training and exposure to professional racing environments with the help of Red Bull's resources and mentorship, which helped him hone his abilities and gain a competitive edge. Thanks to the team's assistance, he went from being a bright-karting prodigy to a genuine contender in the motorsports world.

Alex started competing in international races during this period, participating in several European competitions. These encounters exposed him to many racing cultures and styles, which were invaluable. Competing in several nations improved his ability to swiftly adjust to diverse driving styles and track circumstances, increasing his adaptability as a driver. He developed as an athlete by learning to enjoy the difficulties of competing against various opponents.

His victory in the British Cadet Championship 2012 was one of the turning points in his early career. This accomplishment was evidence of his diligence, hard work, and personal achievement. It confirmed his position as a rising star in the automotive industry and created new prospects. With this victory, Alex established himself in the cutthroat karting world and showed he had what it took to compete at higher levels.

Alex continued to compete at a young age, showcasing exceptional maturity and skill above his years as he moved into Formula Renault and then Formula 4. The complexity of vehicle racing, the significance of aerodynamics, and the increased speeds and technical demands of single-seater racing were among the new hurdles of competing in these championships. But Alex was well-prepared by his early experiences, despite the heightened pressure and rivalry. His rise in motorsport was made possible by his strong foundation in karting and his unwavering determination.

CHAPTER 3: CLIMBING THE JUNIOR RANKS

Struggles and Breakthroughs in Formula Renault

Alex Albon switched to single-seater racing in 2012 after an outstanding career in karting when he joined the Formula Renault championship. This was a big step for him in his racing career because Formula Renault would put his perseverance and tenacity to the test in addition to his driving prowess.

There was a lot of competition at first. Many of his other drivers were seasoned racers with extensive expertise. Alex faced a challenging learning curve as a rookie. Driving a single-seater car involved several technical differences from karting, such as learning the dynamics of downforce, perfecting braking spots, and maximizing tyre performance. Although the early results were less encouraging than planned, every race taught him something new.

Alex faced several obstacles during his rookie season, including performance-related technical problems and inconsistent race results. The stress of competing at a high level while attempting to acclimate to a new environment added to these challenges. Occasionally, he wondered about his ability to compete successfully against drivers with more experience. But instead of letting these difficulties depress him, Alex saw them as opportunities to improve.

Alex committed to intense training and analysis because he was determined to overcome these challenges. He viewed racing films for hours, trying to decide where he could get an advantage and raise his game. He also relied on his crew for input, forming close bonds with mechanics and engineers who could assist him in perfecting his driving technique. As the season went on, the benefits of this dedication to improvement started to show.

In 2014, Alex's diligent efforts started to pay off. In the Eurocup Formula Renault 2.0 series, he managed to land a ride with the highly-rated ART Grand Prix team. As a result of this adjustment, he was given more excellent tools, encouragement, and a more competitive setting. He refined his skills even further by working with a skilled team that has produced many successful drivers.

His first real break came in the Eurocup Formula Renault season, where he started to show signs of promise and improved his consistency. Alex secured his first podium finish at the famed Circuit de Monaco, which catapulted him into the public eye. This feat cemented his place in the series and gave him the self-assurance to aim for even greater success.

Alex showed incredible tenacity as the season went on, challenging for podiums and points in each race. He discovered how to control and use the pressure to fuel his resolve and attention. With the help of his team and his newly found confidence, he was able to move up the rankings, which led to a solid finish in the championship standings.

Alex kept up his excellent work in 2015, participating in the New Zealand Toyota Racing Series and the Eurocup Formula Renault. This demanding schedule tested his ability to adapt and endure various racing conditions, but he was well-prepared thanks to his prior seasons of expertise. His numerous triumphs and steady high-level performance demonstrated the talent that had attracted the attention of motorsport scouts.

Towards the end of his Formula Renault career, Alex Albon had developed from a teenage driver facing difficulties into a competent and competitive racer with the potential to win. The challenges he encountered along the road significantly shaped his mentality and driving style. They instilled in him the value of tenacity, flexibility, and a strong work ethic—skills that would be invaluable as he pursued his Formula 1 career.

Navigating the Competitive World of Formula 3

Alex Albon transitioned to the Formula 3 racing scene in 2015, joining the prestigious ART Grand Prix team following a solid run in the Eurocup Formula Renault championship. This move was noteworthy since, in the eyes of many, Formula 3 is a vital first step for young drivers hoping to qualify for Formula 1. With experienced drivers and up-and-coming talent, the championship offered a competitive atmosphere where victories and disappointments might occur at any given race.

Alex knew that going into the season, he would face tough competition. Numerous well-known drivers, including seasoned racers who had already established themselves in several junior championships, were on the Formula 3 field. This reality raised the stakes, and Alex realized he needed to perform better to have an influence. He was determined to establish himself as a serious competitor.

For Alex, the season's opening races brought a mixed bag of outcomes. Although he showed glimpses of genius in the qualifying sessions, including lightning-fast lap times, his race results were inconsistent. Manipulating the intricate parts of the Formula 3 cars, such as controlling fuel loads and tire strategy, was more difficult than expected. As a rookie, Alex had to contend with the psychological strain of racing at such high speeds against a skilled field, in addition to the technical difficulties.

Despite the difficulties, Alex showed incredible fortitude. He used every race as a teaching moment, evaluating his performance and asking his teammates for input. He greatly benefited from the vital engineering assistance and extensive expertise of the ART Grand Prix, which helped him refine his strategy. The group pushed Alex to adjust swiftly, stressing the value of gaining self-assurance and keeping a competitive spirit despite obstacles.

As the season went on, Alex started to settle into a routine. He improved his ability to interpret the race and gained a more robust understanding of the car's dynamics. During the second half of the season, this growth culminated in a breakthrough. He performed exceptionally at the famed Spa-Francorchamps circuit in Belgium, finishing on the podium after starting from the second row. This accomplishment demonstrated his aptitude and signaled a shift in his driving confidence.

Off the track, Alex encountered difficulties all season long. As a rookie driver figuring out the ins and outs of motorsport, he had to strike a balance between the demands of competitiveness and the business of handling media attention and sponsorship obligations. Learning how to stay focused in the face of these distractions was vital. He emphasized the value of cooperation and communication in handling the competitive and business facets of racing while leaning on his group for assistance.

By the end of the Formula 3 season, Alex Albon emerged as one of the most remarkable talents, securing sixth place in the championship standings. He stands out from many of his colleagues because of his perseverance, flexibility, and capacity for learning from mistakes. His time in Formula 3 gave him the knowledge and skills he needed to tackle the demands of higher classes, including the ultimate move up to Formula 2.

Alex's tenure in Formula 3 cemented his status as a budding talent in the automotive industry. During this crucial time, he learned a great deal about perseverance, strategic thinking, and the value of teamwork. Every race shaped him into a competitive force ready for the next leg of his Formula 1 career and advanced his development as a driver.

CHAPTER 4: FORMULA 2 SUCCESS

Podium Finishes and Career-Defining Wins

As Alex Albon rose through the levels of motorsport, his career was defined by several noteworthy podium finishes and victories that demonstrated his developing abilities. These incidents shaped his competitive driving persona and provided the groundwork for his subsequent Formula 1 triumph.

Alex's participation in the 2016 Eurocup Formula Renault series was one of the critical moments in his early career. He won his maiden race at the storied Monaco track after a season of skill-building. Because it occurred in a renowned competition renowned for its challenging layout and historical significance in motorsport, this victory was very noteworthy. Albon's triumph in Monaco signaled a sea change in his career and demonstrated his ability to go up against the world's top young players. It also gave him priceless experience under duress and putting race plans into action in a high-pressure setting.

In 2018, after achieving success in the Eurocup Formula Renault, Alex advanced to the FIA Formula 2 Championship. Since Formula 1 is frequently considered a direct feeder series for the F2 series, this move proved to be another crucial moment in his career. Albon secured his first F2 podium in the Bahrain Grand Prix, coming in second following a close race with his competitors. His performance in Bahrain cemented his place as a significant contender in the championship by showcasing his ability to adjust to the demands of the car and the opposition.

But Alex started to show her stuff in the second half of the 2018 Formula Two season. He had his breakthrough in the Silverstone feature race, winning with a brilliant drive. Albon showed off his tire management and racecraft skills by starting from the front row and executing a faultless race strategy. In addition to being a personal accomplishment, this Silverstone win sent a message to the motorsport community that he was prepared for the next level.

The Formula 2 season of 2019 saw him maintain his momentum. Albon won his second-season race with another strong performance at the Red Bull Ring in Austria. This race was important because it showed he could execute under duress and pressure. Since Albon was already a member of Red Bull Racing's driver development program, his performance in Austria only strengthened his standing inside the organization.

When Alex was promoted to the Red Bull Racing team mid-season in the 2019 Formula 1 season, it was arguably the most pivotal point in his early career. He debuted his team at the Belgian Grand Prix and placed a creditable fifth. But it was at the next race at Monza that he grabbed attention. Albon battled his way to a second-place finish in a breathtaking show of brilliance, his first-ever podium in Formula 1. This accomplishment demonstrated his ability to compete at the highest level of motorsport and cemented his status as a driver to watch. Finishing on the podium at Monza, a track with a storied past and ardent supporters, was incredibly satisfying.

Throughout his career, Albon's versatility and dedication to constant progress have allowed him to score podium results and victories that have defined it. Every podium finish and triumph has served as a tribute to his talent as well as the dedication and hard work of his teammates and mentors. These moments of triumph have defined Albon's trajectory, which has enabled him to continue pursuing perfection in Formula 1.

Earning Respect in the Motorsport World

As Alex Albon rose through the motorsport ranks, he demonstrated his driving prowess and built a respectable name in the fiercely competitive racing world. Gaining the respect of teammates, supporters, and peers required a steady progression characterized by reliability, professionalism, and a strong work ethic.

Albon's persistent dedication to progress was one of the things that made him worthy of admiration. He was ready to pick things up quickly and adjust from the beginning when he first started karting; these qualities were visible when he competed in various racing classes. Alex frequently conversed with engineers and mentors in Formula Renault and Formula 3 to gain knowledge and improve his abilities. This proactive strategy was not overlooked. Colleagues and team members respected his resolve to improve as a driver because they saw his commitment and work ethic.

In 2018, Albon switched to Formula 2, which was a significant turning point in his quest for recognition in the motorsport world. In this fiercely competitive competition, he competed against some of the world's most promising young talent, including future Formula 1 stars. His play over the season demonstrated his fortitude and capacity to perform well under duress. Albon's deft maneuvering and astute reasoning helped him win the Silverstone feature race and demonstrated his capacity to outrun and outwit more seasoned rivals. This triumph was a significant turning point in his career, explaining his preparedness to compete at the top levels with teams and opponents.

Albon's reputation was also greatly enhanced by his sportsmanship and courteous manner. In a sport where rivalries and competition are expected, he persevered with elegance and humility both on and off the track. Albon maintained his composure and manners during challenging races where he encountered misfortune or arguments with other drivers. His reputation as an honest driver was further cemented by his capacity to face defeats without becoming negative or placing blame.

During his time with Red Bull Racing in 2019, he became well-respected. After being promoted to the team in the middle of the season, Albon had to adjust to a demanding new environment with high standards. It was admirable how quickly he adjusted to the car and how effectively he performed under duress. Albon's incredible second-place performance at the Brazilian Grand Prix, where he started from the fourth row, was one of the highlights. This performance demonstrated his ability to compete against some of the finest in the sport without losing composure; it was more than just about taking home the podium. He gained admiration from his colleagues and rivals thanks to his remarkable racing craft and capacity to stay up with more seasoned teammates.

Albon's cultural heritage also helped him get respect from the motorsports world. As a Thai-British driver, he established himself as a role model for aspiring racers from various backgrounds. Because of his distinct viewpoint and life experiences, he was able to reach a larger audience and win the respect of both drivers and fans. Albon frequently advocated for diversity and inclusivity in motorsports and discussed the value of representation in the industry. His actions in this area struck a chord with many people, strengthening his standing as a well-respected person in the field.

CHAPTER 5: THE CALL TO FORMULA 1

From Formula 2 to Toro Rosso: The Big Break

When Alex Albon joined the Toro Rosso team from Formula 2, it was a significant turning point in his racing career and a chance for many drivers to chase their dream of competing in Formula 1. This was more than just a switch of teams; it resulted from years of arduous effort, tenacity, and the capacity to grasp essential opportunities when they presented themselves.

Albon attracted the attention of Formula 1 teams in 2019 following a stellar season in Formula 2, during which he placed third in the championship with an incredible three wins and numerous podium places. Expectations for his future were raised because of his exceptional racecraft and ability to perform consistently above his peers. Albon's ability to perform well under pressure and his willingness to advance to the next level were critical factors in his success in Formula Two.

Toro Rosso, which would subsequently change its name to AlphaTauri, was a team with a reputation for developing young talent and giving drivers a chance to establish themselves in Formula 1. Due to Albon's affiliation with Red Bull Racing, the team's owner, and his results in the lower formulas, the team decided to sign him for the 2019 campaign. Albon made a significant breakthrough when he was allowed to drive for Toro Rosso, allowing him to demonstrate his abilities on one of the most critical stages in motorsport.

In March 2019, Albon made his Formula 1 debut at the Australian Grand Prix. Fans who had followed his journey were as excited about the event as he was during the weekend. Albon finished in fourteenth place, which was not the outcome he had planned for, but he showed immediately that he could improve and adapt. He made his maiden points finish at the Bahrain Grand Prix, finishing eighth and showcasing the tenacity and will that defined his racing philosophy.

During the German Grand Prix in Hockenheim, Albon had a few good points in his first season. He finished in an excellent fourth place in what would be a pivotal moment of his rookie season, showcasing his ability to maneuver through the challenging rainy weather conditions with a planned drive. He gained respect in the paddock for his ability to remain composed and make wise choices under duress, strengthening his reputation as a talented driver.

Albon's performances improved as the season went on, reaching a breathtaking peak at the Brazilian Grand Prix. After an early incident sent him to the pits, he staged an incredible comeback to finish in second place and secure his maiden Formula 1 podium. This accomplishment demonstrated Albon's capacity to rise above hardship and compete at a high level, in addition to being a personal victory.

From Formula 2 to Toro Rosso, his road was smooth. The challenges of Formula 1 demanded of him a short learning curve and efficient application of knowledge, particularly about the nuances of tire management and car configuration. Albon showed a significant commitment to learning and a strong work ethic; he frequently spent extra time analyzing data and asking his engineers for input. His dedication to professionalism and progress struck a chord with the team, solidifying his standing as a formidable rival.

Albon's ascent through the motorsport ranks exemplifies how important it is to seize the moment in a sport where success frequently depends on timing and opportunity. His move to Toro Rosso allowed him to showcase his skills and established him as a potential candidate for Formula 1 jobs. He would find great value in the skills and knowledge gained during this crucial time in his career as he continued to meet the demands of the top motorsports competitions.

Adapting to the Pace of Formula 1

When Alex Albon joined Toro Rosso for his Formula 1 debut in 2019, he soon realized that the junior formats differ from the Formula 1 world. The unrivaled pace, intensity, and level of competitiveness necessitated extraordinary driving abilities and a thorough comprehension of the automobile and its intricate technologies. Albon's struggle to adjust to this new setting demonstrated his intelligence, tenacity, and will to succeed.

Albon had to quickly get used to the hectic pace of a Formula 1 racing weekend immediately. There needed to be more room for error due to the shortened schedule; he had to learn new material quickly and use it wisely in every session, including practice, qualifying, and the race. Albon had to quickly become proficient with the course layout and car setup during practice sessions. It was imperative to fine-tune the car's performance because even little changes might significantly impact lap times. This adjustment required good driving abilities and the capacity to interact with his technical team in a way that allowed him to comprehend the behavior of the automobile and make recommendations for improvements based on his observations.

Understanding tire management was one of the most critical parts of getting used to Formula 1. In contrast to lower formulas, where tire degradation may be less noticeable, F1 tires need to be maintained with extreme care. Albon had to figure out how to get the most out of the vehicle while keeping the tires at the correct pressure and temperature. His approach evolved to include controlling tire wear, affecting both the race results and his performance. Albon frequently found himself in close quarters with more seasoned drivers during races, with every move affecting his position on the track.

As the season went on, Albon's flexibility grew more and more apparent. He showed signs of becoming more at ease with the car and its capabilities in the mid-season races. His qualifying result was much better, and he could start higher up the grid, which changed his race strategy. Learning from the experience of competing head-to-head against seasoned drivers like Sebastian Vettel and Lewis Hamilton was priceless. Albon became more adept at reading racial dynamics, foreseeing opponents' moves, and knowing how to handle pressure.

Managing the psychological demands of Formula 1 was a crucial component of the adaptation process. Drivers face added strain because of the heavy scrutiny from the media, fans, and team management; Albon was no different. He discovered that he could handle this strain by concentrating on his performance rather than the demands of others. His practice included regular mental training and visualization exercises, which helped him stay calm and focused throughout competitions.

Albon's ability to adjust further demonstrated his eagerness to absorb lessons from errors. Every race presented new obstacles, but he saw every defeat as a teaching moment. For example, Albon spent some time reviewing his team's performance during a difficult race in Monaco, where he had trouble passing and qualifying. He asked for advice on improving his racing lines and handling tight turns. With a proactive mentality, he recovered more quickly and gained points in the next races.

Albon's promotion to Red Bull Racing in the second half of the season was the apex of his adaptation. This transition added another level of expectations and obstacles. After moving to a top team, he had to quickly get used to the subtleties of a more competitive car, where performance was expected to be at the forefront. Albon had to get used to this speed quickly because he was now up against more seasoned drivers and those vying for podium spots.

Albon's confidence increased during the 2020 season as he became comfortable in his new role at Red Bull. He repeatedly proved he could compete at the front, earning significant points and placing on the podium. His experience demonstrated that keeping up with Formula 1's speed is a constant effort that calls for mental toughness, innovative thinking, and a commitment to learning in addition to physical prowess.

CHAPTER 6: RED BULL RACING – THE BIG STEP

Mid-Season Promotion: Joining Red Bull Racing

A turning point in Alex Albon's racing career occurred with his mid-season promotion to Red Bull Racing in 2019. This move demonstrated his talent and the team's faith in his potential. Albon's steady results after a strong debut with Toro Rosso attracted the attention of Red Bull management, leading to a noteworthy career chance that many drivers strive for.

The decision to give Albon a promotion was influenced by several factors. Compared to his teammate Max Verstappen, Pierre Gasly, the driver for Red Bull at the time, had underperformed. The team's management realized they had to make a shift if they would be competitive against rivals like Mercedes and Ferrari. Albon's stellar performances, which included two finishes in which he scored points and a solid fourth-place showing at the German Grand Prix, established him as a viable contender to take on the post.

August 2019 saw Albon promoted to Red Bull Racing, which was a personal accomplishment and evidence of his flexibility and ability to perform well under duress. With his cool head and tactical racing style, he had already shown that he could handle the pressures of Formula 1. Christian Horner, the principal of the Red Bull team, called Albon a "breath of fresh air" and stressed the significance of his elevation in reviving the team's championship run.

Expectations were higher after switching to Red Bull. Albon's decision to join a top team meant that, in sharp contrast to his prior experiences, he would now be racing in a car intended for podium finishes and competitive wins. Red Bull has a strong support network with many tools and knowledge. But that also meant that he had to perform fast, notably when partnered with the well-known and respected Verstappen.

It was a difficult race for Albon to make his Red Bull debut at the Belgian Grand Prix. Even though he came in fifth, the experience taught him the subtleties of driving a more competitive car. To fully utilize the capabilities of the Red Bull RB15, he had to modify how he went. This change was made possible by the vital information that Verstappen and his engineers supplied.

For Albon, the race at Monza in Italy the following year was a turning point. He demonstrated his racing prowess and competitive edge by being in the top six, strengthening his place on the team. His capacity to defend against more seasoned opponents and participate in wheel-to-wheel races demonstrated his increasing competence and confidence. Albon's ability to pick up on and implement the engineers' suggestions soon impressed the crew, allowing him to optimize the car's setup to fit his driving style.

Albon's performances got better as the 2019 campaign went on. At the 2020 Tuscan Grand Prix, he secured his maiden podium result by taking advantage of the chaos during the race to finish in third place. This outcome confirmed that he belonged among the top echelons of Formula 1 drivers and represented a significant turning point in his career. His podium came from skill and poise under duress, which improved his standing in the sport even more.

Albon also established a solid rapport with his players thanks to the mid-season promotion. Working closely with his race engineers to hone his driving and get the most out of the vehicle made communication essential. Albon was a favorite at the Red Bull garage because of his friendly manner and eagerness to learn. His willingness to work together created an atmosphere where receiving feedback was accepted and actively sought, resulting in ongoing development.

Albon encountered difficulties all season long, such as sporadic struggles with a qualifying pace that kept him from starting near the head of the grid. However, his ability to seize opportunities presented a picture of racecraft and smart thought. This ability to bounce back from setbacks became a defining characteristic of his career at Red Bull.

The Pressure of Competing at the Top Level

The strain of competing at the top level of motorsport, especially Formula 1, may be overwhelming for even the most experienced drivers. The move to Red Bull Racing put Alex Albon under even more strain because he was now in the limelight of one of the league's most prominent and competitive teams. The squad, supporters, sponsors, and the media set high standards, creating a thrilling and intimidating environment.

The stakes were evident from the start of his Red Bull race. Albon was joining a well-known squad for its intense rivalry, where the goal was to finish on the podium and earn points constantly. The team's track record of success, particularly under the direction of the highly respected driver Max Verstappen, increased the pressure to produce. Living up to the expectation of being compared to his buddy became Albon's motivation to perform well because he knew they would inevitably occur.

One of the main sources of pressure was the media's and fans' scrutiny. Albon was the subject of extensive media coverage every race weekend, with pundits and experts analyzing every move he made on the track. His performance was closely observed, raising concerns over whether he could maintain Red Bull's Constructors' Championship chances or equal Verstappen's speed. Albon had to retain attention among the din because of the problematic psychological environment created by the continuous evaluation.

For Albon, maintaining his mental health in such a demanding setting became essential. He realized coping techniques were required to manage the pressures of competing at this level. He developed a solid mental game with sports psychologists, using methods like mindfulness and visualization. Thanks to this preparation, he built resilience and the capacity to remain composed under duress, essential for making snap decisions when competing.

The influence of social media constituted another facet of the strain. Albon received both praise and criticism on the internet during a time when fans could interact with drivers directly. Fans' encouraging remarks served as inspiration, yet unfavorable remarks were detrimental. Albon discovered how to get around this digital environment by ignoring unnecessary noise and concentrating on helpful criticism. He underlined how crucial it was to surround oneself with friends and teammates who understood the demands of the game.

Because Formula 1 is a competitive sport, every second matters on the track. Albon frequently found himself up against the clock in addition to his rivals. The pressure to get a decent grid position increased during qualifying since he knew that starting farther up may significantly impact his race. There was a lot of nervousness about not meeting the team's expectations, especially in the critical qualifying sessions where a few milliseconds could mean the difference between an excellent and a poor run.

The unrelenting pace of the F1 calendar increased the pressure. Albon had to swiftly adjust to various courses and conditions because he had back-to-back races in separate countries. Every race had difficulties, from track knowledge to tire management, which necessitated constant learning and adjustment. Albon had to be mentally and physically at the top of his game throughout the season because of the demanding travel schedule and the physical strain of driving at high speeds.

Albon also had to live up to Red Bull's reputation. The squad has a long history of producing accomplished drivers who have won numerous World Championships. Along with that legacy came the unspoken expectation that he would also help the club succeed. Sometimes, this extra layer of strain got to him, especially when things didn't work out as he had hoped. It took some getting used to this expectation, and he had to learn to emphasize his journey more than the legacy of past winners.

Albon found inspiration and moments of happiness despite the difficulties, which kept him going. The excitement of racing, the team spirit, and the backing of his loved ones and supporters provided a solid base. He frequently talked about how these encouraging people helped him handle the strain and reminded him of his original motivation for pursuing a career in racing.

CHAPTER 7: UPS AND DOWNS AT RED BULL

Memorable Races and Near Misses

Alex Albon's life as a driver has been influenced by a combination of thrilling victories and disappointing near misses during his Formula 1 career. These races not only demonstrate his talent and tenacity but also the erratic character of motorsport, where success and failure frequently coincide.

The 2020 Belgian Grand Prix in Spa-Francorchamps, a track known for its challenging layout and erratic weather, was one of Albon's most memorable races. Beginning in the third row of the grid, Albon had to contend with strong opposition from seasoned drivers. He showed great racecraft throughout the race, making daring overtaking moves that put him in the running for a podium finish. During the last few laps, Albon engaged in intense racing against drivers like Charles Leclerc and Sergio Pérez, demonstrating his methodical yet aggressive driving style. He ultimately finished the race in a solid third position, earning a podium for the first time in Formula 1 history and making history.

But only some noteworthy events were worthy of a podium. Albon was given a unique opportunity in the 2020 Turkish Grand Prix. Running in second place for much of the race, Albon found himself in a solid position as the rain-soaked conditions turned the race into a contest of ability and strategy. The track got progressively slicker as the laps went by, which caused many cars to lose their grip and spin out. Albon had to deal with this as well. He battled hard to hold onto his place in the last few circuits. Still, a poor decision in the previous few corners caused a devastating spin, eliminating any possibility of him placing on the podium. With the possibility of a historic finish, he was disappointed to cross the finish line in fourth place.

The 2021 British Grand Prix at Silverstone, a circuit rich in history, was another event that demonstrated Albon's potential. As a member of Red Bull Racing's junior team, Albon had the chance to prove his abilities in front of elite opposition. He engaged in head-to-head racing against Lando Norris and Lewis Hamilton, displaying his expertise and perseverance following a great start. Unfortunately, a frustrating end to a good race occurred when a late-race incident involving Hamilton led to a dramatic collision that took both cars off the course. There was no denying Albon's frustration—he thought he had been unfairly forced off the track in a race that could have earned him valuable points and respect from the paddock.

On the other hand, Albon and the Red Bull team experienced an emotional rollercoaster at the 2021 Hungarian Grand Prix. There were a lot of incidents early in the wild race, but Albon could stay competitive by navigating through the chaos. His team's bright pit plan put him in a solid position to contend for a podium result. But as the race neared its finish line, a late safety car altered the course of events and had teams reconsider their plans. With a respectable sixth-place result, Albon demonstrated his versatility and ability to take advantage of unforeseen events.

These races presented difficulties even though they were terrific. Albon had a lot of near misses that were teaching moments. The Italian Grand Prix at Monza in 2020 was incredibly moving. Albon was in a great position to win his maiden race following a stunning turn of events involving both Mercedes drivers' crashes. But his dreams were shattered when he was forced to retire from the race due to a late-race collision with his former teammate, Pierre Gasly. This incident proved to be a pivotal point in his season, highlighting the narrow margin that separates Formula 1 success from failure.

The Austrian Grand Prix in 2021 was another example of Albon's near-miss incidents. He got off to a strong start in the race, staying in the middle of the pack and appearing ready for a top-five finish. But a string of mishandled tires and hurried pit stops caused him to fall back in the pack, where he eventually finished in seventh place. He was frustrated about losing points, especially since he thought he was moving quickly enough to contend for a better spot.

Facing Criticism and Staying Resilient

Throughout his career in Formula 1, Alex Albon has had his fair share of criticism due to the intense scrutiny surrounding his performance. Because of the enormous stakes involved in the sport, drivers frequently face intense scrutiny from fans and pundits, who magnify every error and analyze every failure. Albon has needed excellent mental toughness and resilience to survive in this hostile environment.

Albon discovered early in his 2019 Red Bull Racing career that success in Formula 1 came with high expectations. The team, well-known for its winning mentality and track record, put tremendous pressure on itself to perform. Albon's performances in his first season combined outstanding victories and discouraging blunders. Even though he had several notable performances—such as finishing third in the Belgian Grand Prix—he occasionally found it difficult to keep up with teammate Max Verstappen's speed. Because of this inconsistency, many fans and commentators questioned if he was the proper player for the squad.

During the 2020 season, criticism grew more intense, especially when Albon encountered difficulties during race weekends. The more Albon's driving style and racecraft were scrutinized, the more Verstappen analogies there were. He was criticized for not having the aggressiveness and decisiveness necessary to succeed on a top squad. Such remarks could easily damage a driver's confidence, but Albon understood that criticism came with the job. He decided to put his attention on his desire to do better and the helpful criticism from his team rather than letting it control him.

Albon grew to see resilience as a critical component of his racing strategy. He knew he had to be psychologically challenging to succeed in such a cutthroat atmosphere. Albon collaborated closely with sports psychologists to create techniques for retaining composure and concentration in adversity. His preparation included techniques like mindfulness and visualization, which allowed him to separate himself from the pressure and focus on giving it his all.

One particularly significant event occurred during the British Grand Prix in 2020 when Albon was in a great position to finish on the podium. He put much effort into holding his position against more seasoned drivers as the race progressed. But Albon ended up in the gravel after a contentious accident caused by a late incident involving Hamilton. Many rebuked him after that race, saying he should have exercised more caution. Rather than giving in to the pressure of popular opinion, Albon used the event as a teaching moment. He reflected on the event, evaluating what he could have done better, and channeled that energy into his next race.

Albon was also under much pressure from the media, who frequently reinforced unfavorable stories about his performances. He used a selective engagement strategy with the media to counter this, emphasizing pleasant contacts and avoiding sensationalist opinion. He ensured that his team members agreed with his aims and tactics by keeping lines of communication open, which contributed to developing a positive work atmosphere even in the face of difficulties.

Moreover, Albon's resiliency was tested by the distinct problems posed by the COVID-19 pandemic. It became imperative to adjust to the new conditions when the season's schedule was changed and races were conducted under stringent health regulations. Albon maintained his concentration on his performance, utilizing his time off the track to study statistics, get fitter, and solidify his racing tactics. This flexibility was crucial in the face of difficulty, demonstrating his dedication to ongoing development.

Albon has seen ups and downs in his career, but his tenacity has eventually transformed criticism into admiration. His achievements started to speak for themselves, and he gained respect for his perseverance and capacity to overcome obstacles. His remarkable skill in races—like his second-place finish in the 2020 Tuscan Grand Prix—silenced his detractors and gave everyone a reminder of his potential.

CHAPTER 8: STEPPING BACK – A YEAR ON THE SIDELINES

Dealing with Disappointment: Losing the Red Bull Seat

One of the most challenging events in Alex Albon's career was losing his place at Red Bull Racing following the 2020 season. The intense pressure of Formula 1 can often lead to the harsh reality of disappointment. This choice especially hurt Albon because he had experienced highs and lows throughout his tenure with the squad and was keen to establish himself on one of the most significant stages in sports.

Albon experienced some turmoil towards the conclusion of the 2020 season. Even though he had demonstrated moments of brilliance, such as multiple strong drives that led to podium results, his overall performance metrics were worse than those of his colleague, Max Verstappen. Albon found it challenging to continuously demonstrate the potential Red Bull expected from their drivers because of his difficulties in qualifying and a few on-track mishaps. As a result, the squad decided to start Sergio Pérez in his place for the 2021 campaign.

For Albon, hearing the news was a huge shock. Losing his seat seemed like a personal defeat because he had joined Red Bull as a promising driver with a solid history. Albon had many feelings following the announcement, including disappointment, despair, and concern about the future. He had dedicated much of his time, effort, and passion to racing for Red Bull. It was a bitter thing to accept, particularly considering how hard he had worked to fit in with the team and how much pressure he had endured all season.

After the announcement, Albon gave himself time to reflect on his adventure. Even though the initial disappointment was unbearable, he tried to deal with his emotions constructively. He tried to contact his loved ones and close friends, who were a huge help during this trying time. Their support helped him remember why he initially chose this path—his passion for racing. He had to remind himself that one failure did not define his entire career and to stay grounded.

Willing himself to remain optimistic, Albon shifted his attention to his alternatives for the future. He was no longer a driver for Red Bull, but he could still be in the Formula One paddock. It was both thrilling and intimidating to think of racing with another team. Albon realized that to attract companies, he had to continue to be proactive and demonstrate his skills. His tenacity was evident when he looked at options to keep him near his favorite sport—a reserve driver position.

Albon joined Red Bull Racing as a reserve driver as the 2021 season went on, allowing him to continue his relationship with the team and prepare for a possible comeback. He also had the opportunity to gain knowledge from the sidelines by witnessing racing tactics, team dynamics, and the regular activities of an F1 team. Knowing that he wanted to be prepared for any chance that might present itself in the future, he made the most of this time to hone his abilities and concentrate on mental and physical fitness.

As Albon processed the disappointment of losing his seat, his moral fiber was tested. He was the subject of media attention and conjecture about his future, but he persisted in his resolve to maintain his composure and concentrate on his objectives. He connected with supporters on social media, comforting him during this trying period. These exchanges reminded him that he still had a long way to go and that many people supported and believed in him.

Albon was given a fresh opportunity in mid-2021 when he signed a contract with the DTM (Deutsche Tourenwagen Masters) to compete in the racing season. By switching to a new series, he was able to rekindle his love of racing and acquire new abilities. Playing in the DTM allowed Albon to show off his skills outside of

Formula One, strengthening his resolve to return to the top of the motorsports world.

Maintaining Focus and Preparing for a Return

After the conclusion of the 2020 campaign, Alex Albon was left without a job at Red Bull Racing, leaving him in a precarious situation. He was disappointed, but he wouldn't let that define his career. Instead, he stuck to his goal of going back to Formula 1 and concentrated on keeping himself mentally and physically ready for any chance that could come around.

Albon started working for Red Bull as a reserve driver in 2021, giving the team vital backroom help. He collaborated closely with the engineers, providing input on simulator sessions and aiding in the development of the cars. He was able to continue learning and maintaining a connection to the Formula One world in this job. Albon embraced this phase with zeal and dedication when fighting on the grid. He knew that keeping his competitive edge and honing his talents would be necessary to make a comeback.

Besides his work on the simulator, Albon followed a strict exercise routine. Because of the extreme physical demands of racing, Formula 1 drivers must maintain optimal physical conditions, and Albon saw the need to be prepared for the race at all times. He put a lot of effort into his training so that he would be ready to take the wheel again if the chance came up.

Apart from his physical preparation, Albon focused on enhancing his mental toughness. The mental strain of losing his position taxed him, as might be expected from competing at the top level of motorsport. But Albon faced this problem head-on, concentrating on his readiness, work ethic, and attitude—the things he could control. During this time, he was able to take stock of his trip, pinpoint areas in need of development, and get ready for new chances.

Albon's commitment was not unappreciated. His drive to improve and good impact on the Red Bull team demonstrated his professionalism and maturity. Even though he didn't compete on race weekends, he remained familiar with the paddock and maintained regular communication with Red Bull's management and team head Christian Horner.

As the 2022 season drew near, Albon's diligence and devotion were apparent. An offer of a full-time ride from Williams Racing signaled his return to the grid. Albon was prepared to take advantage of this opportunity because of his preparation and laser-like focus during his year-long hiatus from racing, and his results with Williams have subsequently validated his status as a formidable competitor in Formula 1. His story shows perseverance through difficult circumstances that can eventually result in success and new chances.

CHAPTER 9: THE COMEBACK WITH WILLIAMS

A New Beginning: Signing with Williams F1

When Alex Albon joined Williams Racing for the 2022 Formula 1 season, his career became better following a difficult time characterized by despair and uncertainty. In addition to being a new beginning for Albon, this transfer allowed him to show off his skills on the track and rekindle his love for racing.

Albon switched to Williams during a pivotal period in his life and career. Albon could stay in the F1 paddock and maintain his talents after leaving Red Bull Racing by accepting a reserve driver position. Still, he was itching to get back into racing competitively and jumped at the chance to work with Williams. Williams, a Formula 1 team with a long history, was going through a rebuilding phase and trying to get back into the race. Albon saw this as more than just another chance to compete; it was also a chance to join a group dedicated to getting better and moving closer to victory.

Albon and the Williams team were thrilled and full of hope when the news of his joining was announced in September 2021. Thanking Williams for the opportunity to get back on the grid, Albon said, "I'm looking forward to this new chapter in my career." I'm thrilled to help this venerable squad with its illustrious past return to the forefront." This upbeat attitude shaped what would be a crucial season for him.

He had to get used to a new squad and the dynamics of the car as he prepared for the 2022 campaign. Williams was going through a transitional period after having had a difficult season. However, Albon significantly contributed to the team because of his F1 experience and comprehension of car setup and performance. His time as a reserve driver and prior involvement with Red Bull gave him the expertise to support Williams in further developing their vehicle.

Albon faced several obstacles at the start of the 2022 season but did so with a strong feeling of will. He got along well with everyone right once and collaborated hard with the engineers to get the most out of the Williams FW44. His early races proved fruitful as he demonstrated his racing prowess and capacity for taking advantage of opportunities, culminating in points-scoring finishes at the Australian Grand Prix. This accomplishment was significant to Albon and Williams since it signaled a positive beginning for their collaboration.

His incredible performance at the 2022 Italian Grand Prix in Monza was one of the highlights of his season. He finished 15th after navigating the wild race with ability and making calculated moves that produced an outstanding outcome. Williams earned its first points of the season as Albon placed tenth at the end of the race. Albon's performance raised the team's spirits and showed his talent and resiliency, demonstrating that he could perform well under trying conditions.

Albon rediscovered his love for racing after joining Williams in the race. He enjoyed driving and racing again, free from the severe pressure surrounding his time at Red Bull. His performances showed this newfound enthusiasm, as he started to run with a confidence that had occasionally been absent from his prior Formula 1 career.

Williams taught us about growth, self-discovery, and racing along the way. Albon enthusiastically jumped at the chance to help the team grow, offering insightful criticism that helped the car improve throughout the season. In the Williams Academy, he also became a mentor for the younger drivers, imparting his wisdom and experiences to assist in developing fresh potential in the field.

Proving Himself Again on the Grid

Following his contract with Williams Racing for the 2022 Formula 1 season, Alex Albon was driven to prove he still possessed the necessary skills to compete at the pinnacle of motorsport. After overcoming obstacles and disappointments in his former Formula 1 career with Red Bull Racing, Albon came into this new chapter with a fresh sense of purpose and dedication. He was resolved to dispel any skepticism about his ability as a race driver and demonstrate his capabilities.

Williams, a club that had endured difficulties in previous years, had varying expectations going into the season. Nevertheless, Albon accepted the challenge of driving for a team going through a change, seeing it as a chance to demonstrate his abilities and collaborate with the engineering team to improve the vehicle. The club could rely on his knowledge and professional demeanor as they worked to move up the standings.

Albon made an immediate impact in the 2022 season's opening races. He demonstrated his tenacity and understanding of racing at the Australian Grand Prix by finishing in the points. This race marked a significant personal achievement for him and enabled Williams to earn his first points in the championship since 2019. Despite his difficulties, this performance demonstrated his will to succeed and his ability to seize opportunities.

A particularly memorable part of Albon's season was the thrilling Italian Grand Prix at Monza. Although he started the race in fifteenth place, he was able to negotiate through a chaotic field that witnessed many incidents and retirements. Albon moved up the grid with a combination of deft driving and calculated decision-making, ending in tenth place and adding another crucial point for Williams. Albon's performance in Monza proved critical to the team's success and development, showcasing his resilience in the face of unforeseen circumstances and pressure.

Albon proved himself during the season by routinely surpassing other drivers and teammates. He frequently found himself amid combat, displaying his forceful but methodical racing style. One noteworthy occasion occurred at the Singapore Grand Prix, where Albon performed admirably, finishing in 11th position despite the problematic street circuit conditions. His ability to get the most out of the vehicle in trying conditions further cemented his standing as a competent driver.

His bond with the Williams group grew stronger as the season continued. He developed into a vital member of the team dynamic, offering insightful criticism that helped shape the FW44. His knowledge and expertise were invaluable to the engineers as they adjusted the car's configuration to improve its competitiveness and performance on the track. Albon's dedication to cooperation and teamwork revealed his development as a driver and an essential member of the Williams household.

Albon's unflinching fortitude in the face of difficulty was one of the most outstanding features of his comeback to racing. After leaving Red Bull, he was met with doubt and criticism, but he took on every task head-on. Albon's performance over the season exemplified his character and a showcase for his driving abilities. He didn't allow the setbacks of the past to control his future; instead, he stayed committed to his objectives.

By the conclusion of the 2022 season, Albon had effectively re-established himself as a competitive driver in Formula 1. He was well-liked in the paddock because of his steady point-scoring and tactical racing style. Albon set out on a quest not only to disprove the opinions of others but also to convince himself that he belonged on the grid. His accomplishments served as a reminder that in the very competitive world of Formula 1, perseverance, hard work, and dedication can lead to redemption.

CHAPTER 10: BEYOND RACING – THE MAN BEHIND THE WHEEL

Exploring Alex Albon's Personal Interests and Hobbies

Beyond the racecourse, Alex Albon is renowned for his wide range of passions and pastimes that provide him with a fulfilling existence outside of Formula 1. Although racing takes up a lot of his time and attention, Albon's hobbies and interests outside of driving reflect his lively nature and help him stay healthy overall.

Albon's love of cooking is one of his most prominent pastimes. He finds inspiration and comfort in experimenting in the kitchen. Cooking gives him a therapeutic release and a way to decompress from the high pressure of racing. Albon frequently posts pictures of his culinary masterpieces on social media, featuring everything from contemporary fusion recipes to classic Thai delicacies. His mother is Thai. Therefore, this link to cooking relates to his history and demonstrates his understanding of the culinary traditions of other countries. His passion for cooking gives his life a feeling of routine and facilitates interactions with loved ones over shared meals.

Apart from his passion for cooking, Albon also enjoys taking pictures. He often takes pictures of his travels, particularly on race weekends, with intense energy and excitement. Albon's ability to recognize the beauty of the world outside of the racecourse is demonstrated by the numerous beautiful photos of the towns and landscapes he sees that appear on his Instagram account. In addition to showcasing his artistic side, his photography endeavors give followers access to the locations and events that motivate him away from the racecourse.

Albon also enjoys playing video games, and he does it frequently in his own time. He enjoys playing video games as a great way to relax and spend time with friends, both locally and virtually. Many athletes now enjoy playing video games to escape the mental strain of training and competition and to compete in friendly matches. Albon has competed in several esports competitions, displaying his sense of competition in a new setting. In addition to providing pleasure, his gaming passion helps him form bonds with friends and fellow drivers with similar interests.

Albon's love of animals, especially cats, which he frequently highlights in his social media posts, is another facet of his private life. His pet gives him comfort and companionship, especially on busy racing weekends when the stress of competition can be too much to handle. Albon loves the little things in life that make him happy, even in the competitive world of Formula 1. His love demonstrates this for animals.

His passion for health and fitness is also essential to him. To stay in top physical shape, he engages in a variety of exercises, including cycling and gym sessions, in addition to the physical demands of racing. He regularly encourages his fans to prioritize their health by sharing his fitness journey. His dedication to physical health is also reflected in his support of mental well-being, as he sees the value of leading a balanced life incorporating mindfulness exercises like yoga and meditation. This all-encompassing approach to health and wellbeing keeps him resilient and focused in the intensely competitive world of Formula 1.

Albon has also stated that he is interested in motorsports outside driving. He likes interacting on social media with his fans and followers, and he has even taken part in charitable activities to promote motorsport and its beneficial effects on local communities. His desire to interact with fans and give back shows that he recognizes the importance of fostering racing for the next generation of fans and the broader impact that racing has on society.

Honoring His Thai Heritage

Thai ancestry has greatly influenced Alex Albon's identity, morals, and way of living, both on and off the track. Albon was born in London to a British father, Nigel, and a Thai mother, Kankana. Albon has balanced the two cultures throughout his life, finding pride and inspiration in his Thai heritage. His personality is based on this cultural dichotomy, which shapes his outlook and strengthens his ties to Thailand and the UK.

Albon's love of Thai food is among the most apparent ways in which he pays tribute to his Thai ancestry. He was exposed to traditional foods as a child, many of which he now likes to prepare and serve to teammates and friends. His favorite foods, such as spicy papaya salad and pad thai, reflect his mother's cuisine and the flavors he has always loved. Albon frequently posts about how much he appreciates these dinners on social media, demonstrating his culinary prowess and strong bond with his Thai heritage in a world where fast racing rules; sharing these experiences helps to preserve and honor his history.

Beyond just food, Albon is deeply rooted in his culture and actively participates in Thai society, upholding its customs. Albon participates in family-oriented, communal, and spiritual celebrations during major Thai festivals, like Songkran, which is the traditional New Year's Day observed in April. He frequently lets his followers in on these festivities, giving them a peek into the colorful culture that helps to define his personality. Albon promotes an increased understanding of Thai culture within the international motorsport community by openly upholding these traditions.

Albon keeps a close relationship with his family in Thailand and enjoys their food customs and holidays. He travels the nation frequently, taking part in philanthropic endeavors and events. Albon has a particular enthusiasm for assisting disadvantaged Thai youngsters and local communities. His charitable endeavors, motivated by a desire to give back to the community that has shaped him, demonstrate that he carries the spirit of his heritage with him wherever he goes.

He also recognizes the difficulties of driving a mixed-race vehicle in a field that Western and European drivers have historically influenced. He has made speeches on the value of diversity and representation in motorsport. Albon encourages others to embrace their heritage and follow their aspirations, regardless of their cultural or ethnic identities, by sharing his experience as a Thai-British driver. His participation in Formula 1 shows the value of diversity in the sport and the strength of inclusivity.

Albon's ties to Thailand are further evidenced by the encouragement he gets from Thai supporters. He has a sizable fan base in Thailand, and his followers greatly appreciate his accomplishments on the racecourse. This fervent fan base serves as both a source of inspiration for him to perform and a reminder of his resolve to represent Thailand internationally. He frequently flaunts his national identity and wears a helmet with the Thai flag as a mark of his fidelity and commitment to his roots.

His ties to his heritage also influence Albon's attitude to cooperation and teamwork. The values of humility, respect, and community are highly valued in Thai society. Albon is constantly willing to learn and adjust, which is visible in his relationships with his colleagues. His demeanor of deference towards his peers and rivals reflects the fundamental principles his Thai background instilled in him.

CHAPTER 11: LESSONS IN RESILIENCE

Staying Positive in the Face of Adversity

Alex Albon has encountered many difficulties in his career that have tested his fortitude and tenacity. Sustaining an optimistic mindset in the face of hardship has been a defining characteristic of his personality since his early karting days and his experiences in Formula 1. Albon's story serves as an example of how having a positive outlook may promote achievement and personal development in addition to assisting in overcoming setbacks.

When Albon was in Formula 2, he faced one of the biggest obstacles imaginable. He was pressured to perform as a rookie driver, especially considering the high standards his teams and sponsors set. His debut season was a disaster, a string of unlucky events that cast doubt on his ability to compete in motorsports. Albon decided to concentrate on what each race had to teach him rather than give in to prejudice. He sought input from his coworkers and engineers, carefully examined his performance, and put forth endless effort to improve. His capacity to maintain positivity and an open mind helped him hone his abilities and ultimately produce outstanding outcomes, culminating in a third-place position in the championship rankings.

Albon's move to Formula 1 was another challenging time that tested his determination. He soon discovered that the competitiveness at the highest level of motorsport was unlike anything he had ever encountered after being promoted to Toro Rosso in 2019. There was a lot of pressure to produce results immediately, especially given the history of the drivers who came before him. Albon accepted the challenge and saw every race as a chance to improve. He frequently discussed the value of maintaining optimism and emphasized how having a positive outlook enabled him to negotiate the challenges of Formula One racing successfully. This way of thinking made him better and created a positive team atmosphere where everyone was welcome to participate and work together.

When Albon was promoted to Red Bull Racing in the middle of the 2020 season, it was one of the most pivotal events in his career. A high learning curve was associated with the shift and greater scrutiny and expectations. Albon experienced several setbacks, such as challenging races and backlash from the public and media. He was under increasing strain, yet he remained upbeat and saw every failure as an opportunity to grow and learn. Albon frequently expressed his conviction that, in these difficult circumstances, it was essential to remain committed to his objectives and have faith in his skills. Several outstanding performances, including powerful drives that showcased his ability and potential, showed this resiliency.

Albon's upbringing also influences how he handles hardship. His upbringing in a multicultural environment taught him the value of tenacity and flexibility. As a child, he developed his ability to negotiate many cultural contexts, which taught him to be adaptable and upbeat in the face of difficulties. These events formed his perspective on the world, and he gained the mental fortitude to confront challenges head-on.

He has frequently acknowledged that his support system keeps him upbeat in addition to his competitive spirit. During difficult moments, his teammates, friends, and family are vital sources of support and perspective. Albon often thanks the people in his life, realizing their encouragement was essential to his achievement. He frequently stresses how crucial it is to surround oneself with uplifting people and build a solid support network because doing so can help one overcome difficulties.

Albon has also used social media to interact with and motivate those going through difficult times by sharing his experience with followers. By candidly sharing his challenges and the lessons he has learned, he has created a sense of community among followers who can identify with his experiences. His willingness to be genuine and vulnerable speaks to many people, showing that growth and resilience can come from having a good outlook and that facing challenges is acceptable.

The Importance of Patience and Hard Work

In motorsport, skill alone is frequently insufficient for success; a driver's ability to combine patience and hard effort sets them apart. These two characteristics have been essential to Alex Albon's racing career, helping to mold him from a youthful karting enthusiast to a powerful Formula 1 competitor. His experiences exemplify how developing patience and putting forth constant work may result in long-term success.

Albon stood out from his colleagues early in his racing career because of his excellent work ethic. He didn't rely only on natural skill when he started karting but also dedicated himself to mastering the sport's nuances. This involved learning the ins and outs of his kart, honing his driving techniques, and researching his rivals. Albon would frequently practice for hours at the track, carefully evaluating his performance and looking for ways to improve. This commitment demonstrated the value of perseverance in accomplishing objectives and set a strong basis for his future undertakings.

Albon faced more pressure and competitiveness as he moved to higher-level competitions, such as Formula Renault and Formula 3. The value of patience became apparent in these demanding situations. Many obstacles had to be overcome before success, such as depressing race outcomes and periods of self-doubt. Albon, however, discovered that these experiences could be seen as instructive rather than sad. His understanding that every setback presented an opportunity for development strengthened his resolve to put in the required effort.

Even though the short-term outcomes were unfavorable, he focused on his long-term objectives because of his patient attitude.

His journey to Formula 1 served as another example of the value of perseverance and diligence. In 2019, he secured a position with Toro Rosso following a prosperous tenure in Formula 2. It wasn't a smooth transfer to Formula One, though. It's a massive step from a junior series to the top motorsport division, and Albon had a long learning curve. He had to adjust to the complexities of a more complicated vehicle, the challenging racing tactics, and the fierce rivalry between the world's top drivers. Rather than hurrying to establish his worth, he deliberately emphasized steady progress over instant recognition. This viewpoint relieved some of the burden and allowed him to approach every race comfortably.

Albon's unwavering work ethic is among his career's most notable traits. His devotion to maintaining physical and mental well-being and his technical grasp of the vehicle are evidence of this. Albon frequently stresses the value of training off the track, which involves mental activities and physical training to improve concentration and decision-making during competitions. His strict training reflects his conviction that perseverance is the key to success, which supports the notion that innate ability must be developed via steady effort.

Additionally, Albon's tenure with Red Bull Racing emphasized the value of patience in his professional life. He was under tremendous pressure to perform well after receiving a promotion in the middle of the season. Fans, the media, and his team had high expectations for him. Albon knew it would take patience and persistence to prove himself. He encountered both thrilling victories and vexing setbacks throughout this stage. He showed his potential as a competent driver by remaining patient and keeping up the hard work. His capacity to rise to the occasion under duress was demonstrated by his excellent performances, which included a podium finish, which was the consequence of his determination.

Outside of the racing world of competition, Albon's work ethic and patience have found application in his personal life. He is well-liked by fans and other drivers because of his modesty and commitment to ongoing growth. Albon frequently discusses the value of upholding a solid work ethic and a patient mindset, stressing that hard work, perseverance, and a willingness to learn are all factors that contribute to success rather than just skill.

CHAPTER 12: LOOKING FORWARD – THE FUTURE OF ALEX ALBON

Hopes and Goals for His Formula 1 Career

As he continues to forge his path in Formula 1, Alex Albon's aspirations and objectives reflect his enduring desire and distinct future plans for the sport. Albon's journey has been filled with setbacks and victories, and as a result, he has a clear idea of what he wants to accomplish at the highest level of motorsport.

One of Albon's main goals is securing podium positions and triumphs in Formula 1. He knows the joy of battling at the top of the grid, having felt the rush of standing on the podium during his time with Red Bull Racing. Albon stated that he wanted to build on his early success and would put much effort into improving every race. His time at Williams F1 has strengthened this objective as he looks to increase the team's competitiveness and show he can be a front-runner in the sport.

Apart from attaining personal triumph, Albon is deeply conscious of the significance of collaboration in Formula 1. He wants to build relationships with engineers and other drivers by promoting a cooperative and upbeat work environment within his team. Albon understands that the driver alone won't guarantee success in Formula One; the driver and the team must work together for success. He hopes to improve communication and planning during races by developing a close relationship with his crew, resulting in better performance. His emphasis on teamwork stems from his conviction that obstacles can be more successfully addressed when presented as a united front.

Albon also hopes to serve as an inspiration for young drivers, especially those from different backgrounds. As a product of a cosmopolitan upbringing and a lack of diversity in the sport, he aspires to motivate the next generation of racers. He wants to inspire young drivers that success is attainable with perseverance and arduous effort, no matter what background they come from, by sharing his story and experiences. Albon's wish to serve as an inspiration demonstrates his dedication to advancing diversity in the motorsports industry.

Albon has also indicated a desire to help Formula 1 develop, notably in sustainability and technical innovation. He wants to contribute to the sport's shift towards cleaner methods and electric power, pushing for advancements that improve efficiency and environmental consciousness. Because of his engineering background and technological interest, he is a suitable advocate for these improvements, and he sees F1 leading the way in sustainable motorsport in the future.

Regarding his objectives, Albon wants to keep improving as a driver. He knows that the world of Formula 1 is changing quickly and that flexibility is essential. He is dedicated to constant self-improvement through demanding training and performance analysis to remain competitive. Albon watches race films on a daily basis to learn more about his driving technique and pinpoint areas that need improvement. His development depends on this analytical approach, which he uses to improve his technique and reach his full potential on the track.

Albon is also acutely aware of how critical it is to keep an optimistic outlook as his profession develops. Given the intense pressures of Formula 1, he wants to develop resilience to overcome obstacles. By emphasizing mental toughness, he hopes to stay focused on his objectives while navigating the highs and lows of racing. This viewpoint keeps him grounded and maintains his motivation despite obstacles.

The Legacy He's Building in Motorsport

While Alex Albon pursues his Formula 1 career, he considers his accomplishments and the legacy he is leaving behind in the motorsport world. Albon knows the value of making a lasting impression beyond his racing career, as shown in his dedication to excel while making a meaningful contribution to the sport.

Albon leaves behind a lasting legacy as a pioneer of diversity in racing. Due to his Thai and British ancestry, he represents multicultural representation in a sport where drivers from the West have long held a dominant position. Albon's accomplishments show that brilliance can arise anywhere and open doors for ambitious racers from various backgrounds. His prominence in Formula 1 conveys to young athletes that hard work and enthusiasm can overcome obstacles. He hopes to inspire the next generation by being transparent about his upbringing and experiences and encouraging them to follow their aspirations despite their challenges.

Albon is also dedicated to encouraging inclusion in the sport. He proactively interacts with supporters and other drivers, promoting a culture that values diversity and invites involvement from all groups. His goal in speaking at events and participating in outreach activities is to raise awareness of the difficulties experienced by under-represented groups in racing. Albon's initiatives ensure that motorsports develop into a more welcoming environment for all people by fostering a more significant conversation about diversity.

His incredible performance-related recovery after losing his seat at Red Bull Racing says volumes about his tenacity and willpower. Rather than letting failures define him, he used the chance to get back up, improve, and establish his value. In addition to showcasing his driving prowess, his outstanding accomplishments with Williams F1 also showed his capacity for adaptation and success under trying conditions. Aspiring drivers can draw inspiration from this perseverance-filled voyage, which supports the notion that achievement is achievable even in the face of difficulty.

Albon also strongly commits to motorsports' future and sustainability. He understands how critical it is to participate in the industry's shift towards greener technologies and practices. Albon's support of ecologically conscious racing indicates a growing consciousness among drivers regarding their environmental impact. He advocates for sustainability to positively impact the sport's future development and ensure it remains relevant and responsible. His participation in talks on the direction racing technology is taking shows that he wants to make a lasting impact on environmental stewardship and performance.

Beyond the mechanics of racing, Albon cherishes the relationships he has made throughout his career. He has developed relationships with teams, fans, and other drivers, highlighting the value of friendship in the motorsports community. These connections foster a community that facilitates mentoring and knowledge exchange. Albon's readiness to lend a hand and share his experiences strengthens the sense of community in motorsport. This supports the notion that success is frequently the result of a team effort rather than an individual endeavor.

CONCLUSION

Alex Albon's motorsport career is proof of the strength of tenacity, willpower, and grit. From his early years traversing the unique cultures of Thailand and the UK to his ascent through the Formula 1 rank, Albon has encountered many obstacles that have molded him into the driver he is today and the person he wants to be. His story is one of conquering challenges and the never-ending pursuit of greatness; it is more than just a run of races and podiums.

Albon recognizes the value of each experience, highs, and lows, as he reflects on his career. He has gained priceless knowledge about strategy, collaboration, and self-belief from each race. He hasn't let the obstacles he faces, such as the pressure of taking the Red Bull Racing position and the heartbreak of losing it, stop him. Instead, they have strengthened his determination to advance and be successful. Albon's resilience and dedication to his work are demonstrated by his capacity to stay goal-oriented in the face of difficulty.

He has proven during his path that he is not just a talented racer but also can motivate others. He appreciates his role as a role model for aspiring drivers from varied backgrounds and is aware of the importance of representation in motorsports. Many people who have experienced hardships can relate to his story, emphasizing the value of perseverance, hard effort, and the guts to follow one's aspirations.

Albon is enthusiastic about the prospects ahead of him and the future. He is still setting high standards to become a dominant force in Formula 1 and significantly impact the sport. Because of his experiences, he now has a profound respect for the trip itself. He understands that success is shaped not only by the accomplishments but also by the connections made and the legacy left behind.

To ensure that both facets of his identity are honored in his job, Albon is committed to upholding his Thai ancestry while embracing his British roots. His continued dedication to diversity and sustainability in the motorsports industry demonstrates a deeper awareness of the obligations of being a public person in the modern world. By supporting these causes, Albon hopes to create a lasting legacy that goes beyond his accomplishments on the track.

Alex Albon's motorsport career story is one of resiliency and victory. His narrative encourages others to follow their passions, overcome obstacles, and persevere in chasing their goals. Albon is a living example of the idea that every chapter in the life of a motorsports enthusiast adds to a more extraordinary, more significant legacy and that the trip holds equal significance to the final goal.

www.ingramcontent.com/pod-product-compliance
Lightning Source LLC
Chambersburg PA
CBHW050317230526
45471CB00005B/2232